BEYOND THE STARS

BEYOND THE STARS

Discoveries in Astrophysics

ROWAN EVERHART

RWG Publishing

CONTENTS

1	Introduction	1
2	The Birth of Astrophysics	3
3	The Expanding Universe	4
4	Stellar Evolution	5
5	Black Holes and Neutron Stars	6
6	Exoplanets and Planetary Systems	8
7	The Search for Extraterrestrial Life	10
8	The Mysteries of Dark Matter	12
9	The Nature of Dark Energy	14
10	The Multiverse Theory	16
11	Gravitational Waves: Ripples in Spacetime	18
12	The Role of Supermassive Black Holes	20
13	The Formation of Galaxies	22
14	The Life and Death of Stars	23
15	The Evolution of the Universe	25
16	The Origins of Cosmic Rays	27
17	The Cosmic Microwave Background Radiation	29

18	The Structure of the Universe	31
19	The Interstellar Medium	33
20	The Role of Magnetic Fields in Astrophysics	34
21	The Impact of Supernovae	36
22	The Study of Gamma-Ray Bursts	38
23	The Role of High-Energy Particles in Astrophysics	40
24	The Search for Dark Matter Particles	42
25	The Study of Extragalactic Astronomy	44
26	The Role of Observational Techniques in Astrophysics	46
27	The Future of Space Telescopes	48
28	The Role of Computational Astrophysics	50
29	The Challenges of Astrobiology	52
30	The Role of Astrochemistry in Understanding the Universe	54
31	The Study of Stellar Populations	56
32	Conclusion	58

Copyright © 2024 by Rowan Everhart

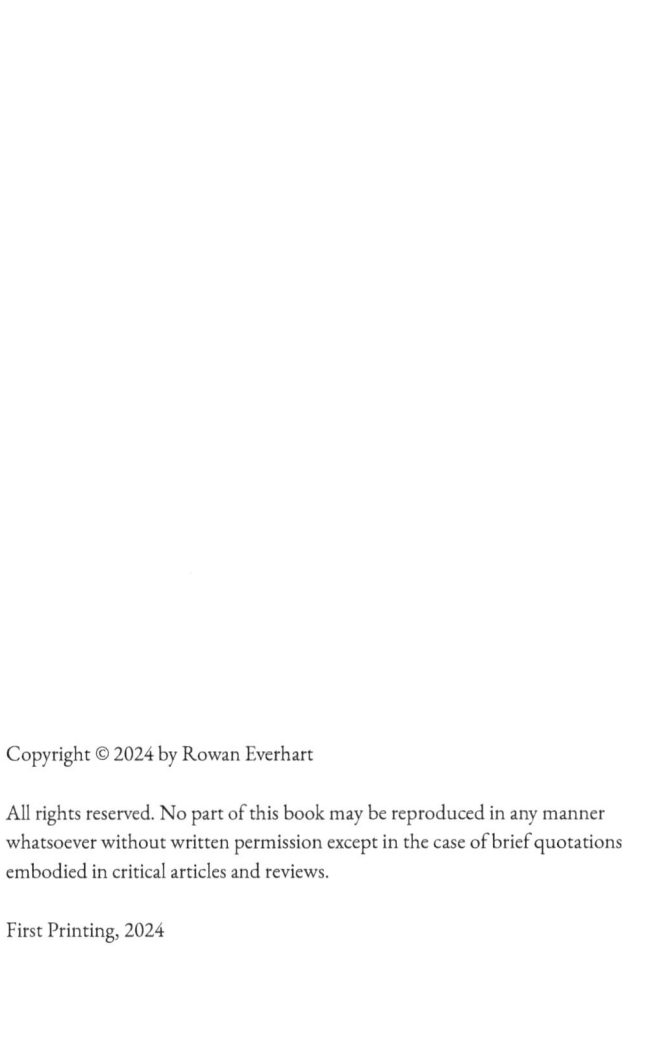

All rights reserved. No part of this book may be reproduced in any manner whatsoever without written permission except in the case of brief quotations embodied in critical articles and reviews.

First Printing, 2024

CHAPTER 1

Introduction

Quasars, while the true powerhouses over a two-decade span between mid-seventies and early nineties, did their exhaustive flaring very deep in the most distant reaches of the cosmos, and for a reason not having to do with supernovas. The evidence was that their activity and indeed their existence required extreme conditions. The first quasars contained enormous amounts of highly ionized iron, which requires a close strong ultraviolet radiation field to exist locally. Very soon, the same sources were found to resemble but to exceed in power local active galaxies, and these are indeed abundant in many such galaxies. And when radio galaxies in any large sample became known, they too had their quasarean boosts. As though the host contained a dark, gravitationally associated monster, the rapidly spinning accretion of whose mass onto a central black hole that lit up the whole structure with its radiant power, reaching out to the edge of the immense halo of catatonic iron-containing vapor whence it poured. The Ashker model predicted the required rapidly spinning brightness and there was good reason to believe that the spin might have been entered at birth.

More than three decades ago, a renaissance of astrophysics began, inspired by a telescopic search for black holes using the then newly discovered phenomena of cosmic X-rays. Astronomical objects overwhelmingly declared themselves not only to be marvelously intriguing in terms of the physical phenomena they contain (which for many years has been the characteristic of astrophysics), but also sources of complex outputs of star formation, of energy budgets in both radiation and in nonradiative forms, and of industry in the processing and dissemination of chemical products. We plan to discuss two domains in which astrophysical studies, sought broadest but inevitably tiniest of evidence to test the most general of general understanding, have paid off, and delivered intriguing new puzzles because of how they differ from what is nevertheless otherwise well-established physics. The evidence for the puzzles (and the puzzles) interact inasmuch as each is a hint which constrains the other either by a discrepancy or by a lack thereof. Even so, both domains are sufficiently controversial, still, that one might rightly conclude that the subjects are worth much further experiment, which in each case we shall then discuss.

CHAPTER 2

The Birth of Astrophysics

However, Galileo was certainly not the first to make use of the telescope for astronomical observations. It wasn't long before other celestial objects were discovered, such as the four largest moons of Jupiter (which are therefore known as the Galilean satellites). These satellites provided vital observational support for the idea that the Earth circles the Sun, as proposed by Copernicus, rather than all other celestial bodies circling the Earth, as claimed by the approved cosmology of the Catholic Church and by its chief philosophical defender, Aristotle.

The birth of astrophysics is inextricably linked with the invention of the telescope by Galileo in 1609. This allowed observations of the heavens to be recorded with better spatial resolution - and, crucially, to be repeated and improved - than the naked eye could ever manage. His most famous observations of the heavens are also those whose consequences for the history of ideas had the most far-reaching impact: the discovery of the satellites of Jupiter and the phases of Venus. These demonstrated that celestial bodies orbit, or are orbited by, something other than the Earth.

CHAPTER 3

The Expanding Universe

In 1927, the Belgian astronomer-physicist Georges Lemaître visited Lemaitre's idea was that the universe began with a hot explosion, now known as "The Big Bang", proposed by his colleague and friend Hermann. Lemaitre's most significant contribution to the expanding universe came when Hubble discovered that the universe was not only expanding but also getting larger with time. That implies that if we look to distant galaxies, they are flying away from us at higher speeds as he observed. Hubble's results were described in the now famous Hubble's Law, and Lemitre's theoretical work on what we now call the expansion of the universe was based on the solutions of Einstein equations proposed previously by Friedmann but went unnoticed.

The discovery that the universe is expanding came as a great surprise. For a long time, it was thought that the universe was not changing with time. In 1917, Albert Einstein added a term to his famous equations for gravity. Without his cosmological constant, the equations predicted that the universe should either expand or collapse. To keep the universe static, under the pressure of his contemporaries who believed that it was unchanging, Einstein had to introduce this arbitrary constant. Just two years later, Russian astrophysicist Aleksandr Fridman showed that the Einstein equations actually predicted a universe that would expand. However, his work remained almost unnoticed till decades later.

CHAPTER 4

Stellar Evolution

Invited review to appear in: Proceedings of "Beyond the Stars: Discoveries in Astrophysics" - XXXth Reunión Bienal de la Real Sociedad Española de Física, Madrid, Spain, September 13-17, 2002. The interest in the formation of neutron stars dates back to the first detection of a radio-frequency signal originating from the nearby supernova remnant Cassiopeia A. This signal, detected in 1968, is thought to originate in an ionized shell surrounding the putative weakly interacting compact object that remained after the supernova explosion of a massive star just 337 years before, in 1631.

Although the life cycle of low-mass stars is determined by a process dubbed the "initial mass function," the sequence of events that could occur at different stages of such a life cycle leads to the conclusion that low-mass stars are probably the main contributors to the enrichment of the interstellar medium. Neutron star formation is studied as the result of the protoneutron star radiation. The entropy growth associated with such high-energy photon radiation leads the iron core to bounce out and further collapse into a neutron star. The possible transition from the proposed phase of superluminality to photon viscosity is analyzed. Closed-form solutions for the transport along a neutrino-dominated neutron star envelope are presented.

CHAPTER 5

Black Holes and Neutron Stars

When the observed mass distribution of stars is combined with the expected supernova masses, it is predicted that neutron stars and black holes will be formed in a roughly equal number of events. If a significant amount of the carbon and oxygen cores of collapsing stars are not able to collapse directly into black holes, can this be critically tested using astronomical observations? The answer is yes. Black hole formation, and especially the mass limits of black hole formation in nature, may have a critical effect on massive binary evolution. This event was interpreted as a stellar-mass black hole being formed by the direct collapse of a massive star's core, since the mass of the black hole is estimated to be more than four times the mass of the sun, which is thought to be the maximum mass of a supernova remnant neutron star. It is important to note that black holes can also be formed through other processes in the universe, such as primordial black holes left from the early universe, but the mass limits of black hole formation and the process of their formation are completely different from black hole formation.

Black holes are regions of spacetime that have so much warping of the fabric of spacetime that no matter or radiation can escape them. They can form from the remnants of massive stars that undergo core collapse at the end of their lives. When a massive star exhausts its nuclear fuel, it will eventually undergo a core collapse and form a protoneutron star. Neutrinos are produced and carry away energy that cannot be held by the protoneutron star. That causes the protoneutron star to lose pressure support and gravitationally collapse to a black hole—typically milliseconds to seconds after the onset of the collapse of the proto-neutron star. Two important factors must be considered for massive stars whose cores are still made of carbon and oxygen (hesitating to go to even heavier elements). The first is that, in the absence of significant mass ejection from the system during the core collapse, the remnant formed via the supernova explosive event is expected to have a mass of 2-3 times the mass of the sun.

CHAPTER 6

Exoplanets and Planetary Systems

The planets of the Solar System were discovered through visible light observations made with telescopes. The first exoplanets, however, were discovered using the radial velocity method. This technique measures the velocity of the exoplanet's host star by observing slight blue shifts and red shifts in the star's spectrum as the star wobbles about the center of mass of the planet and star system. The size of this wobble is dependent upon the mass of the planet, the mass of the star, and the radius of the orbit of the planet in a way that was first described by Johannes Kepler, and it is why the velocity method provides primarily a minimum mass for the exoplanet. In fact, if the orbit of an exoplanet is seen edge-on from Earth, the wobble can be detected even if the planet's mass is much less than the observational sensitivity of the method. To date, almost 90% of all exoplanets have been detected by the radial velocity method.

Published online in 2018, the first exoplanets were discovered during the early 1990s, and the number of detections has been increasing on a nearly exponential curve ever since. There are many exoplanet surveys in the works or already underway. There are also missions such as TESS and Pluto, which are aimed specifically at discovering exoplanets. Because of this, the subfield of exoplanet science is an incredibly fast-moving one. Here, however, we can provide an overview of some of the most important methods and findings from this burgeoning branch of astrophysics.

CHAPTER 7

The Search for Extraterrestrial Life

The search for life beyond Earth provides compelling motivation for the field of exoplanet research. Significant strides have been made since the discovery of the first planet around a solar-like star, 51 Peg b, a 'hot Jupiter' in a three-day orbit. Almost 4000 exoplanets have been found so far and we now know that planets exist in a wide range of masses and sizes that fall outside of our solar system, some like those of our system, and others that are entirely dissimilar. This knowledge pathway has been driven by countless ground-based telescope surveys, followed by the precise radial velocity measurements conducted from the interfaces of our high-resolution spectrographs with the largest telescopes in the world. The Wildcats planet search team, a collaboration between scientists from the Université of Montréal and Institut de recherche sur les exoplanets (iREx) at the University of Toronto, has played a significant role in this search by conducting some of the most successful radial velocity surveys to date. We then expanded our study to include planets from a photometric method, a method that allows us to measure much more significant planets orbits and system compositions through the use of Tomography.

Are we alone in the universe? Addressing this question scientifically requires the identification of extrasolar planets with the right characteristics that allow life to thrive. Once a potential host is found, we must verify if conditions are amenable to life. The key factors for hosting life – water-rich planets, orbiting within the habitable zone and around the right kind of star – require knowledge that extends beyond the scope of the human imagination. The search has only just begun. From the start, the scientific approach has been to learn from the Earth and search for systems that could potentially be 'Earth-like'.

CHAPTER 8

The Mysteries of Dark Matter

The density of light element produced in the very early universe has only recently been deduced. Perhaps the hitherto unseen dark matter is bright, consisting of baryons which contribute to the dark matter of cyclic models by being locked up in low-mass galaxy-like objects or subgalactic clumps produced by intricate non-linear fragmentation. It may be that the uncomplicated dark matter renormalization group transformation breakdown of the sterile neutrinos with mass order of 1 keV, i.e. well below that normally to be associated with the cold dark matter threshold, allows such a particle to be the dark matter. The needed space for inflation must also likely have GUT-like interactions. These show up of enormous interest to the study Planck satellite that aims to elevate the primordial texture CMBR anisotropy signals above the standard inflationary ones for some regions of the texture parameter space and the 1st year WMAP angular power spectrum. Recent publications also discuss the range of parameters associated with UDM models (including the integrated Sachs-Wolfe imprint of DE).

"Had I been present at the creation, I would have given some useful hints for the better ordering of the universe." Thus said the Greek philosopher and scientist Anaxagoras in the 5th century B.C. It is our privilege to live in a time in which significant discoveries are being made that take us back very near the time of the creation with extraordinary clarity and vividness. We are discovering the fundamental laws that govern the workings of the universe we live in, and we are learning something about its origins. We will model and perhaps understand what happens in the first instants of the birth of the universe. We will know about the dark matter that controls the dynamics of the universe. Discovery is our contact with creation. We are pushing the envelopes of space and of time. We are bound to make mistakes, but also progress. As long as what we are searching for are interesting questions that are not completely divorced from the ways that we search for answers, we should expect to learn something, as indeed we do. Astrophysics in the next century will be based on a level of understanding fundamentally different from what it is today. The overall structure of the universe's cosmic microwave background radiation (CMBR) inhomogeneities, for instance, will differ strongly if the universe has inflated by approximately e100, rather than if atoms were indeed the smallest building blocks of matter. The universe's current contents have been probed by recent measurements.

CHAPTER 9

The Nature of Dark Energy

How peculiar that something as unknown as dark energy could be 70% of the mass-energy in the universe. We visualize it as a simple hump in our equations. We need an "i" for a bit of mathematical courage and write this most simple hump with an "i-" in front of it. When we solve these equations, it turns out that the mass-energy density within this point is such that the force it generates causes the cosmic expansion to accelerate. How remarkable! So simple! A natural explanation for a surprising observation! Of course, when we solve the same equations for the aether, or quenched quarks or whatever analogs are proposed elsewhere in the world, they accelerate cosmic expansion too. Nature has played here.

The first two chapters of this book have explained the cosmic history of the universe, taking you from the very beginning to the present day. Throughout that history, it is clear that the universe has nothing if full of surprises. My colleagues and I are driven by our awe at these findings, and we are motivated by a simple human desire to explain them. What drives the astrophysical community to continue to explore primeval fireballs, Big Bang echoes - the cosmic microwave background? We still seek to understand why the sky is dark and occasionally bright. We wonder what lies within the vacuum of space. We wonder about the mysterious clumps of dark matter that dot the universe. These puzzles push us to keep experimenting, to keep simulating, and in a mathematical landscape full of infinities - equations that we write down in an attempt to capture the workings of the universe and then expand and expand until the unwieldy behemoth makes no sense - they keep that most human of traits alive.

CHAPTER 10

The Multiverse Theory

If spatial correlations exist between substructures of our universe and those observable parts of the multiverse, and there is evidence that the cosmic microwave background might have such a deep spatial correlation between the points of appearance of matter, it would provide further constraints.

Circle deviations of comparisons between future measurements of cosmic radiation from clusters with those of the distribution of galaxies within our own cosmic horizon will provide massive constraints on inflation models and also conditions to apply a variant of the gambling methodology to the multiverse and simplify the calculation of the probabilities of emergence of life forms in it.

In the bubble formation model, new universes form at the interfaces of bubbles, so correlations may exist between neighboring regions of the multiverse. Causal regions could cluster anew in the appearance of bubbles. New galaxies are frequently constructed within larger structures formed from smaller galaxies. Long-range observations of galaxies are sensitive to this.

The most acknowledged model proposes a foam-like distribution where the universes are bubbles, only the interior of which become universes. The ones that form before inflation ends are unsuitable for life, as these develop structures too quickly and the first atoms cannot form.

Their cosmological regularities could prevent star formation or lead to the rapid collapse of any formed stars and produce forms of emission that prevent the existence of life forms anywhere.

According to the concept of the multiverse, an indefinitely large system of universes is independent of one another and all occurred from the same initial singularity. These universes will evolve differently, some with galaxies, others with stars and galaxies, yet others will contain nothing of the sort, as will be the case of our universe once it becomes uninhabitable by life forms.

CHAPTER 11

Gravitational Waves: Ripples in Spacetime

Instead, to be detected, gravitational waves must come from cosmic cataclysms. Einstein realized that one such event would arise from two contracting masses. As the ellipsoidal masses orbited each other, Einstein realized that energy must be lost because energy-carrying waves were sent out in the form of gravitational radiation. And as the orbit decayed under the influence of the emitted waves, the two masses would collide. The magnitude of the waves would increase rapidly once the two masses were close together. When initially observed in 2015, the LIGO signal demonstrated that Einstein was right in every detail. This was Albert Einstein's final triumph, for his idea had been realized 100 years later, almost as he had envisioned it.

Gravitational waves: Ripples in spacetime. On 14 September 2015, the great dream of Albert Einstein was fulfilled. At 9:50:45 a.m. Central European Summer Time, gravitational waves given off when two black holes merged about 1.3 billion years ago reached the Earth and were detected by man's first gravitational wave observatory, LIGO. Einstein predicted the existence of gravitational waves in 1915 as part of his general theory of relativity. In his theory, space and time are merged into one entity, spacetime, which is distorted by the presence of matter and energy. Einstein's theory predicts that any accelerating mass must emit gravitational waves. But because gravity is the weakest force, the waves would never be strong enough to produce detectable effects, no matter how large the mass acceleration.

CHAPTER 12

The Role of Supermassive Black Holes

Besides being a powerful consumer of ionized gas, which surrounds them, active galaxies show X-ray emission from an optically-thin iron K-shell line at 6.4 keV and a corresponding iron K-edge at 7.1 keV shifted towards higher energy when the black hole is rapidly rotating. These features enable us to probe the extreme environment of the accretion disk at an atomic scale close to the black hole with the assumption that the X-rays reflecting medium extends down to the ISCO (Innermost Stable Circular Orbit). A further important feature emerging from this model is the presence of a Compton reflection halo with a broad band spectrum peaking in the MeV region. Accurate measurement of this relativistic effect for Mcs 2, the dimensionless black hole spin, opens exciting improvements in the measurement of this fundamental parameter for supermassive black holes. But these measurements are a little biased by the strong gravitational field near the black hole. Naively, it seems that the Fe Kα line and the reflection continuum can be affected by strong Doppler and gravitational redshifts that, in turn, make them good probes of Mcs 2.

Our galaxy, and probably all large galaxies, contain a supermassive rotating black hole at its center. The fact that this central object is a black hole has been recently confirmed by high-resolution observation in the radio and X-ray bands. Accretion of gas and stars onto these supermassive black holes is an important energy source in the Universe, especially for active large galaxies, in which the gravitational energy of infalling matter is partly converted to powerful jets. The usual relativistic jet model is a collimated, supersonic, relativistic outflow from the accretion disk surrounding the black hole thanks to the magnetic field anchored in the disk. Together with its effect on the interstellar medium, such jets are suspected to substantially reduce the angular momentum of matter which is transported to the central singularity.

CHAPTER 13

The Formation of Galaxies

While star formation is an exciting process, stars need the protection of a Sun-like cocoon to form, like the water of a 'Bénard cell' needs Kelly's fingers. However, in 17 my residing in big temperature differences, larger than average, are as much a chance fluctuation for the uf of Bénard cells on a heat interface as for the first grain of dust in an underdense region in order to grow. The condition for such an object, which is called in the language of cosmologists 'protoglobal self-gravitating collapse', imposes no such restrictions.

The amazing thing is that the galaxies which appear on your holiday photographs are rather complex objects made of trillions of stars. Just like in the postal system, where letters pile up in many thousand post offices in many thousand towns around the world, forming a structure of towns, cities, countries, and islands of service, we expect that the larger structures in the Universe are built up by the gravitational attraction of small matter constituents. In the beginning, matter was spread smoothly, but locally, due to small fluctuations and differences in the initial matter density, 'overdense' regions start to evolve faster than 'underdense' regions. Around matter centres, much matter aggregates, and so after a while, groupings of high matter densities form, some of which contain enough matter to start 'cosmological fires', i.e. objects, and so the first star-forming regions emerge.

CHAPTER 14

The Life and Death of Stars

In the denser winds of larger evolving main sequence stars, if the primary stars are close enough together, they will transfer matter into the primary atmosphere which may lift the donor envelope for RX J0806.3+1527 safely above the Roche Lobe. These stars produce buckets of the modified material, consisting of the solar shaped magnetopause, flung into the solar wind. But these longer-lived interacting binary systems also produce very high bulb densities from naked solids. Optical lightflash binaries produce density fluctuations with periods on the order of the last-gasp twinkling period. These sources are pumping energy into neutral media in front of the source and into the companion's wind, at a rate requiring their rapid decay if they are to avoid explosive nuclear burning of surface hydrogen, or indeed to avoid the rapid dissolution of the star in favour of symbiotic activity.

Red angels in the sun – the mystery of proto-planetary nebulas – are nebulous stellar funnels which focus sun-like star winds into two jets which then blast through the stellar cocoon and debris in resonant spheres of glowing debris trailing behind each dusty conical shell close to the star. They are examples of astrophysical pinwheeling, spinning plasmas and pinching magnetic focusing of solar coronal-point plasmas at speeds well below light velocity, another example of Maxwell's fluid analytical dynamics at work in hydro-magnetism. In the circulatory powering of the solar jetting of electrons in the conical outflow typically produce kilo-gaus strength B-fields near the snowflake poles, also producing the high hollow bullet pressure which remains central to the porous scaffolding of specialized heating cells revealed by the solar cycle.

CHAPTER 15

The Evolution of the Universe

On the contrary, supernovae and the passage of comets were images from celestial phenomena. Supernovae were interpreted in several ways: they could be considered to be the result of the conjunction of celestial luminous bodies or were born along the zodiacal circle (according to a somewhat materialist point of view) or close to the Earth (if it is in our power to prevent). The first we now call false phenomena (since they are due to the conjunction of Jupiter and Saturn). Comets were considered to be a kind of planetae, eventually wandering along the Zodiacal circle and appearing as a consequence of Earth in their trajectories. So, it was clear for no one that supernovae were extremely powerful cosmic explosions and that they, as well as the phenomenon of comets, could be used for some physical investigations and, to some measure, for the understanding of the universe.

In the Homocentric models considered by Aristotle, all planets revolve around the Earth. This implies that beyond the sphere of the stars, the order of the various spheres representing each planet is fixed. However, at this time, only six spheres were needed by astronomers to account for the motions of the five known planets. The ancients wondered why God bothered to create an extra empty and useless sphere.

Four hundred years ago, in the lifetime of some of the present astrophysicists, scientific knowledge taught that the universe was composed of a single star surrounded by seven planets and that the Earth was at its center. These were the ideas of Ptolemy, which had dominated Western thought about our place in space and time for fifteen centuries. There was scattered and limited evidence for small deviations from Ptolemy's view. For example, the planets, and in particular Mars, sometimes went into retrograde motion, and also some of the planets appeared to be stationary at opposition. However, this was very different from observational data that required the Earth and all planets to move on ellipticals. It was a simple exercise for Kepler to explain very accurately the observed data. Thus, stellar and planetary observations seemed not to offer promising avenues to abandon the geocentric view.

CHAPTER 16

The Origins of Cosmic Rays

The unexpected acceleration effects which had to be taken into account were described in terms of reflection conditions at a strong collisionless shock. Further theoretical advances were made in 1978 when V.L. Berezinsky, Z. Yoshida, and V.S. Ptuskin demonstrated that the second half of the previous century's progress in cosmic ray observations might find a natural explanation in a diffusive shock acceleration mechanism by amplification of Alfvén waves. It was observed that the limit on the maximum energy, which the shock wave could transport, decreases as TeV particles interact with the thermal plasma of relativistic electrons.

The fact that the mean mass of cosmic rays is ~ 10 g cm-2 in spite of this dilepton energy loss from discrete interactions and the ~ 5 g cm-2 expected energy loss to radiation damping and ionization is remarkable for an interaction-poorly characterized cosmic-ray environment. Several astrophysical shock fronts have locally observed strongly accelerated cosmic ray protons but are or should be inefficient cosmic ray sources due to the very steep power-law behavior of their gamma-ray spectra.

A.C. Janse van Rensburg proposed that these element-dependent effects could be expected if a similar overabundance existed above the mass 4 isotope, and H.J. Volk and I. Gold assumed that these were protons which were spawned by iron primaries. It took about 20 years before the new generation of cosmic-ray physics experiments, notably at the Rockefeller Institute of G.L. and T. Cupo at the University of Naples, would demonstrate that the position of the knee was very different for different elements. This made Volk and Gold's iron-proton chain hard to realize.

That cosmic rays originated in far-off outer space was first established around 1912. To this day, cosmic rays are believed to sample the universe on a scale well in excess of the present observability limit. Excesses in cosmic-ray abundances for specific elements had already been observed in the 1940s by investigations associated with the search for missing matter in the universe.

CHAPTER 17

The Cosmic Microwave Background Radiation

Nevertheless, the exploration of cosmic radiation moved towards the low-frequency region due to advances in engineering, and the property of angular resolution with high sensitivity was also allowed. This technique led to the detection of the anisotropy of the cosmic background radiation. As explained in a previous section, tiny density perturbations entered the horizon during the decoupling epoch (when the first atoms combined). The gravitational potential wells captured the electrons and pushed themselves strongly coupled to the photons: in the collapse, the temperature dropped and the photons were not able to escape from the potential wells. These are characterized by the apparent size of the horizon at that time. Therefore, for perturbations larger than the apparent size of the horizon, the plasma distribution was random and then the perturbations can produce temperature fluctuations in different regions of the sky. At the collapse, the photons of the related distance could escape and so the fluctuations in the temperature of the radiation should have the ability to emerge. This phenomenon is known as the decrease of the photons.

The experiment went unnoticed for a very long time. Then, in the 1950s, scientists carried out the ambitious experiment of listening to the 'music of the spheres', assuming that the Big Bang would have provided an initial intense hum of radiation throughout the Universe. They identified the expected radiation profile, accounting for the expansion of the Universe, but also a mysterious background of unmasked hum that was independent of the frequency and consistent with terrestrial radiation. Symbolically, the Earth was under the radiation shadow of the decoupling epoch and it couldn't have been in a better location.

CHAPTER 18

The Structure of the Universe

The book shows the vast panorama and the first shoot of our human comprehension of the various phenomena at work in nature. It suggests the synthesis and harmony that the human intellect could reach in describing the Universe. The book covers a wide range of topics which are articulated in four parts: the first, on basic physics, gives the main elements for the structure of matter; the second surveys astrophysics and cosmology; the third part mainly focuses on the microworld, and the fourth discusses the most speculative theories, such as superstrings. The second part provides insights on relativistic cosmology and its concepts, from space-time curvature to the basic cosmological parameters, temperature anisotropy, density parameters, cosmic abundances of BBN light elements, and cosmological constraints related to the Hubble expansion, amplitude of the primordial matter fluctuations, high values of age, and deceleration of the cosmological dynamics.

"The Structure of the Universe" provides an exciting and comprehensive view of the cosmos. This textbook offers the general physics leader working in the various fields of astrophysics and cosmology the necessary mathematical, physical, and astrophysical background to understand the structure, history, and the essential features of the Universe as we see it today. One of the major features of the book is that some sections - such as those giving the ultimate theories of physics - are independent of the surveyed astrophysical and cosmological topics provided in the specialized sections. For instance, here we find two chapters with the basics of relativistic cosmology, and the most recent and important achievements of particle physics in unification theories: inflation, topological defects, baryogenesis and magnetic monopoles, and the problem of vacuum symmetry breaking, scale invariance, and superstrings.

CHAPTER 19

The Interstellar Medium

Interstellar gas heats up, and its density increases, through any sort of process that drives atoms forcefully (i.e., at supersonic speeds) into each other - such as supernova shock waves, turbulence driven by, e.g., the spiral density wave of a galaxy, or massive stellar winds. Compressive heating is ubiquitous - the interstellar medium (ISM) is hot! Most of the heating in molecular gas occurs through an energy cascade whereby large-scale structures (caused, perhaps, by supernovae in the parent molecular cloud) drive increasingly smaller and smaller structures that transfer the (turbulence) energy of large scales to the smallest scale of the local molecular gas itself, where it is steeped to heat.

The interstellar medium: Gas is the stuff of stars. There is an enormous amount of gas in galaxies - perhaps 80% of the "normal" (baryonic) mass. Much (perhaps most) of it is very tenuous indeed - Voyager II found much of the area of space between the Sun and the edge of the solar system to be decidedly non-vacuous, as does the dramatic shape of the heliopause from the shape of the magnetic field of the Sun. Even where gas is apparently absent, like the lower density regions surrounding galaxies, gas is present: it just might be very under-ionized.

CHAPTER 20

The Role of Magnetic Fields in Astrophysics

Jupiter takes a few minutes to turn about its axis, so the thin current sheet or magnetotail of Jupiter's magnetosphere derived from the solar wind is distorted and spirals around the planet. Since some of the many moons of Jupiter are maintained in radiation belts similar to those of the Earth, and Jupiter's auroras are particularly intense, they might well emit energy in the x-ray region. To bring the total output more in line with intensity emitted in the ultraviolet regions or in the optical requires some consideration of the role of the magnetic field in the energy budget of the auroral zones. Observations in the x-rays and gamma rays are still quite limited. The reader will soon realize that there is much need for added information. Certainly, energetic particle measurements at high rates are not real imaging of the constellation of x-ray sources of the entire region.

Planetary fields are generated by convective motions in the metal cores. The field of Neptune is largely dipolar. Like the Earth's field, direct observations reveal a reversal associated with the motion of the magnetic equator. With a dipole of B8 T, the planet Neptune is an extremely powerful magnet. Uranus is encased in clouds which hide its surface, where the field is generated. Jupiter is the most powerful magnetic dynamo in the solar system. The field is very nearly a dipole. As the dipole is tilted a few degrees from the rotation axis, the magnetosphere of the radioactive planet Callisto is strongly dipolarized and plasma is in the magnetosheath. This has been observed in situ by the Galileo spacecraft.

In considering the role of magnetic fields in astrophysics, it is useful to have some knowledge of the types of magnetic fields that are observed in the universe. Solar magnetic fields are quite unique because they have been studied in great detail. But very fortunately for astrophysics, that level of investigation has already begun.

CHAPTER 21

The Impact of Supernovae

Beyond the facts enunciated so far, the observation of this single supernova in the Magellanic Cloud led to no revolution in astrophysics. Why then is supernova 1987A given the title of this talk? A similar question was asked about the relatively ordinary and energetic supernova 1054. The answer to that question is provided by oriental archaeology. The crab nebula, product of the explosion of supernova 1054, was depicted over 500 years ago by Chinese, or maybe Korean, or maybe Japanese, astrologers, with an eye for white dwarfs in 1054. Perhaps, this portrait is the first pictorial representation of an object in the universe outside the solar system. It is by introducing a change in the characterization of the physical world by men, a modification of the way by which nature is comprehended, that a cosmic phenomenon, be it the brightness of a celestial body or properties of its light, can be known as cosmological.

In 1987, a brilliant point of light was seen in the southern sky in a region that appeared to have few stars. This new cosmic guest that suddenly appeared in the dark Magellanic Cloud, a small nearby galaxy, was a supernova explosion. Indeed, this supernova shone more than a whole galaxy, outblazing its celestial neighborhood for several months. In the beginning, supernova 1987A shed light on the neutrinos, ghostly particles that travel unimpeded through space at almost the speed of light, penetrating all matter. Those neutrinos had been detected for the very first time in coincidence with the light from a supernova. In later times, the light from this supernova was pried on model-dependent cosmological estimates. Nowadays, high-precision measurements of cosmic parameters damp rather than inflame the enthusiasm of the cosmologists.

CHAPTER 22

The Study of Gamma-Ray Bursts

Gamma-ray bursts for the longest time remained unconfirmed sources, their nature not even agreed upon, except that they could not be closer than say a dozen Mpc due to a lack of sharp parallax of their arrival direction. This well-motivated far away-localization indicated isotropic γ-ray emission energies in excess of 10^51 J. On February 28th, 1997 then, the first distance measurement using optical spectroscopy was made. Located in a faint galaxy at a redshift z = 0.835, the burst could be placed 0.72 Gpc (2.3 light-days) away and the emission energy was measured to be 2*10^53 J (inferred isotropically), casting doubt on all classical models, which were anchored in published datasets that could still solve the implicit, additional model parameters. It also implied that these events dominate cosmic nucleo-synthesis, that they are beamed and that a large rate is inferred when standard candle template light curves such as R(utt-50ms)^-0.5 are assumed (R is peak brightness).

On March 5, 1979, two gamma-ray detectors on the Compton Gamma Ray Observatory recorded a gamma-ray burst that for eight seconds swamped the saturated detectors, lasting 0.2 seconds in the observer frame and energizing the γ-ray photons with a total energy greater than 10^{12} J. Gamma-ray bursts hence became one of the most enigmatic objects in astrophysics.

The study of the most powerful explosions in the universe has experienced rapid progress in the fledgling field of high-energy astrophysics. Due to new theorizing and a renaissance in both the technology of sensors and new theory, rapid progress has taken place, which has led to an unprecedented understanding of what these objects are. In this article, the basic observations that are part of the picture describing gamma-ray bursts will be reviewed to give an overview of constraints. Traditional theory will then be discussed briefly as will the advance in understanding that has taken place in the last 10 or so years. We begin with a review of the observations.

CHAPTER 23

The Role of High-Energy Particles in Astrophysics

Secondly, even if cosmic-ray acceleration as we know it will not work, the high-energy particle component may still be relevant astrophysically. There are, of course, numerous objections to the above conclusions at the same time when the object which we label, the relativistic component, are by far the strongest intervening field in the Universe - are destroyed rapidly, all the way from cool Jeans-unstable protoclusters with characteristic magnetic fields of 10's of uG, to interstellar voids with even stronger B-fields, before the age of even the youngest hic-mesocores begin to exert their 'influence' in the interstellar medium. Further, at late evolutionary times, after supernova-driven shells begin to overlap, the density of the confined material even exceeds its initial value - it has been "swept up" by the pressure waves emanating from both the forward and reverse shocks. Only a confined, relativistic component provides sufficient pressure to drive this expansion. In fact, a hydromagnetic logarithmic structure is required to provide sufficient feedback on the ionized shell at all times higher or lower. However, in the absence of a concentrated momentum-dependent magnetic field - a steady-state interstellar accretion flow is extremely unlikely, especially if the mass of the central object is not overwhelmingly larger than the mass in the galaxy.

In the latter part of the 1990s, a number of exciting papers were published, relating the acceleration of particles in supernova remnants to observations both at high energies of cosmic rays and at lower energies which are not affected by the extremely high galactic magnetic fields and are therefore "faithful", optically thin representations of the original cosmic-ray sources. As described during this conference, numerous nearby molecular clouds possess an energetic ionized boundary shell which have many properties in common with the predictions of collisionless particle acceleration, i.e., no matter how realistic our models, exhibit remarkable astrophysical phenomena at low values of the plasma-parameter, and even in situations as diverse as close, colliding, massive stars - or - a warm plasma, permeated by magnetic fields, a highly conductive medium. It was also pointed out that magnetic fields would, in all situations of practical interest, be completely unaffected by the relativistic particle component.

CHAPTER 24

The Search for Dark Matter Particles

Even though direct and indirect detection of neutralinos face many experimental and astrophysical problems, the possible detection of some γ-ray lines in the galactic halo may be one of the best ways to confirm the hypothesis that weakly interacting supersymmetric particles are responsible for the formation of cosmic structures. These annihilation lines, in fact, are subjected only to astrophysical uncertainties which can be evaluated within a factor of two to three, which is much less than in the case of models that give only a continuum photon spectrum.

The accurate description of the expansion of the universe is of major importance for our understanding of cosmic evolution, black holes and wormholes, galaxy formation, and the formation of large-scale structures in the universe. Among other contributions, the discovery of the cosmic microwave background radiation has confirmed our understanding of the initial conditions for the formation of cosmic structures. Today, the numerical computations of the time evolution of the universe assume the existence of a dominant component named cold dark matter, which interacts only gravitationally with the other known components. Neutrinos are, in fact, one of the best dark matter candidates, still, they constitute only 20 to 25% of the matter required for a match between theory and observation. It is therefore important, especially because of the very small mass which the Higgs mechanism bestows to them too, to test other ideas about the nature and the properties of dark matter particles.

CHAPTER 25

The Study of Extragalactic Astronomy

For astronomers, one of the most important problems has been to identify the distant and faint extension of our Milky Way, called the island universe, that we now call galaxies. Also, we would like to study various phenomena relating to galaxies. In fact, cosmic rays mainly consist of protons that pour onto us from all directions, not only from our galaxy but also from the entire universe. Although they originate from various sources, such as supernova explosions and black holes present only in our galaxy, about 90% of them originate in powerful launching jets that reach enormous energy up to the entire universe from active galaxies at the center of some galaxies. These accelerated protons collide with other protons and various materials in the interstellar space and are thus disintegrated into particles in a complex process, causing the emission of electromagnetic radiation, mainly gamma-rays.

Since antiquity, humans have lived under the stars from birth to death, helped by the stars in various ways. At the same time, people have been treating astronomy as a great spiritual phenomenon. Perhaps these were the reasons why so many people, including me, dream of becoming astronomers. In fact, however, over the past 70 or so years, most astronomers have turned their attention from the study of individual stars in our galaxy to the investigation of various objects and phenomena that occur in this universe, from the period immediately after its birth down to recent years in an ever-increasing area.

CHAPTER 26

The Role of Observational Techniques in Astrophysics

The method of optical spectroscopy is widely used. With the development of CCDs, the size or position in the sky of an object is determined by automatic center-finding algorithms and with standard star isophotal magnitudes calibration, which means that it is generally no longer necessary to prepare a reference image and measure brightnesses within a particular aperture. For faint objects, density reconstruction is quite important in any form of confusion-limited image analysis, such as isoplanatic correction techniques. While there is considerable power in the Fourier techniques, the large interpolations in tracer data will lead to a significant level of noise in the real space correlation function measurements, systemic bias in the angular correlation function, and potentially a non-negligible contribution to the errors in the cosmological parameters. To some extent, this interpretation of the imaging data may be generic, but, since our primary goal is to concentrate on insights to be gained from the real and Fourier space aperture photometry for isoplanatic filled aperture imaging on spatial mass fluctuations, it is useful to explore why redshift space information can be advantageous and what information is lost.

Until a few centuries ago, observations of the sky were made with the naked eye. Changes in the sky were recorded, as were the positions of the stars. With the advent of optics in the 17th century, measurements of the positions of stars could be made with great accuracy. In the 19th century, Bessel measured the first parallax angle of a star and Carte du Ciel photographed the sky to great precision. Astrophysical observations began properly in the early part of the 19th century, with the identification and classification of the many new astronomical phenomena that the telescope brought to the fore. With the discovery of the new celestial objects, new observational techniques came into play, sending back information on the phenomena in question, such as telescopes with spectral analysis, and photographic plates with the prism, coordinates, and colors. It is these two methods that briefly illuminate the different procedures in astrophysics.

CHAPTER 27

The Future of Space Telescopes

Hubble will be complemented by the James Webb Space Telescope (JWST) and the proposed Wide Field Infrared Survey Telescope-Astrophysics Focused Mission (WFIRST) and its planned successor WFIRST-AFM-2, as well as additional missions that provide cosmological information, such as CMB-space and the Large Synoptic Survey Telescope (LSST) over similar time scales. These missions will continue the legacy and success we have in detailing and explaining cosmic structure, but practical considerations about the U.S. state and project funding could push other missions even farther into the future, asking: What other astrophysical science can space telescopes achieve that go beyond what those planned space telescopes can do?

Experiments the Size of Galaxies would permit us to study tiny variations across entire galaxies and measure the geographic distribution of dark energy and dark matter, to measure the statistical clustering of different galaxy populations over vast volumes, and to measure how galaxies evolve on a truly cosmic scale. Experiments that map structure when our Universe was only 200 million years old could detect the most massive black holes that are the seeds of early galaxies, find the birth pangs of star formation, understand the role of dark matter and cosmic background radiation fluctuations of structure growth, and detect a broad range of relativistic effects relevant to understanding the very early Universe.

Goal: A general review of the science that will be performed with the James Webb Space Telescope (JWST) and the proposed Wide Field Infrared Survey Telescope-Astrophysics Focused Mission (WFIRST-AFM) after 2024. Those missions will enable new astrophysical science but also will inform engineering for the next generation of general-purpose telescopes.

CHAPTER 28

The Role of Computational Astrophysics

The purpose of this volume should be to increase our knowledge in this matter. Renowned workers in those and adjacent disciplines agreed to participate. Full agreement will not result, but we can as the holy cow of previous discussions. While several contributors have reviewed previous works and results. Papers on the Sun, the Compact Extragalactic Radio Sources, and gamma-ray bursters are welcomed for the reason they represent the limit of our knowledge. Those results are, of course, included in this volume, with reference to follow-up studies by those contributors and other researchers.

The importance of computational astrophysics in modern astrophysics cannot be overstated. It is just not possible to trace every step of every process from the fundamental laws and to solve the resulting system automatically to reach the same results. Being able to spend the necessary resources and time, and the feasible processing units to affect them with, we need approximate methods of solving astrophysical processes. Whenever possible, accurate but numerically limited calculations are to be preferred over approximate methods of a problem. Full accuracy apart, however, there is wide agreement in non-local methods, by which new or improved physics can be included without the need of endless modifications. The predictive ambitions must, of course, be checked against the use of, and could in non-local calculations. Objective of this volume, which collects results of studies mostly were of less of the same character, with positive and negative outcome in recent years. Clearly, star formation and evolutionary processes are another part which uses a combination of local and non-local methods. The problems related to this subject and the issues of chemical evolution have been extensively investigated, often combining analytical and numerical experience.

CHAPTER 29

The Challenges of Astrobiology

That naturally brings us to the field of astrobiology, to the study of life in the Universe. This has always been a topic of interest, but has tended to be labeled as philosophy or pseudoscience. As Robert Richardson so aptly pointed out in his excellent book, Many Worlds, it is in a sense a topic that makes sense to treat philosophically, that borders on the meaning of life. But since the exploration of Mars and the moons of the outer planets, which will come to fruition in the next decade or two, the question of life beyond the Earth has gained a new respectability. Bioastronomy, or the Search for Extraterrestrial Life, has become a proper topic for our scientific community. There is too much we do not know and yet could know in the next decades on this topic and its potential effects on the Human Condition, to dismiss it lightly."

"Just as the discovery of the electron preceded the development of electronic technology by four decades, scientific discoveries of the nature and magnitude now on the horizon will bring forth profound changes in our way of life - in our technology, in our exploration, in our mental and medical outlook - in the next century. Unlike on previous occasions, when the results of man's artistic and mental endeavors have generally had a positive effect on his destiny, the fruits of scientific research are ambiguous and may or may not have a benign effect on mankind. We are extremely clever on developing our technology, but whether we are coping adequately with its beneficial and sinister effects is a serious question.

CHAPTER 30

The Role of Astrochemistry in Understanding the Universe

Molecules appear under the effect of photons, neutral as well as ionizing, which generate ionization but also break the links formed by the chemical species on the bases. The formation of molecules was possible according to the available data about the chemistry of molecules, ions, radicals, and then complex molecules, but also having in mind the ionizing radiation emitted by the first stars which were massive, the most abundant emitted radiation with wavelengths less than those. The first stars thus play an essential role in the formation of molecules, one of the essential keys to the appearance of life, as the elements necessary for life were formed by nucleosynthesis processes which took place in these first objects.

Astrochemistry has found the complex and valiant exercise of detailing the chemical network that describes the reactions that are possible when molecules, ions, atoms, radicals, and photons are in the astrophysical environment. But how were the first molecules formed in the universe as we know it? When were they formed? What are the keys to the emergence of ever more complex molecules? We propose some responses from the results (laboratory experiments, theoretical calculations, but also astronomical considerations including observations of deep space as well as those closer to us, in the solar system).

CHAPTER 31

The Study of Stellar Populations

In a general relativistic cosmology, especially one with a large cosmological constant and a recombination redshift of about 1000, and perhaps even in an open universe, the stellar populations left behind at that time can interact with their radiation, if the radiation can escape into the void around them, to set up regions of baryon pressure and of radiation pressure. These will slow down the onset of structure formation. The dark matter, however, does not interact with the radiation, and can already have formed small collapsed masses, which then provide the acceleration field necessary to accelerate the baryons to fall into these potential wells. Part of the collapsed structures of dark matter, the baryons, and the radiation then slowly move in the direction of mutual gravitation, to form one ensemble of identity of purpose. Only a short time later, at some redshift between 1000 and 1600, does this ensemble of identity of purpose form anything recognizable as a specific galaxy.

Stellar populations are the ensembles of stars at the present epoch that have come into being at discrete epochs in the past. Such populations command our attention for several reasons. As our understanding of the redshift-distance relation, also called the Hubble flow, improves, we become more and more confident that all galaxies have commenced their careers as concentrations of stars, rather than as gaseous regions.

Therefore, for the astronomer interested in the Universe at very early times, the place to search is in the originating stellar populations. We may also be concerned with the life processes of stars such as nucleosynthesis, and with high energy phenomena such as gamma ray bursts that accompany the formation of some compact objects, if not all compact objects. But nowadays the crowded environments of galaxies are the scene of many high-energy phenomena, and bring about interactions beneficial to the realization of complex molecules.

CHAPTER 32

Conclusion

From the point of view of instrumental and conceptual methods of Physics, Astrophysics has developed and validated many. A general idea is that we cannot directly manipulate and test the majority of the physical phenomena that happen in the Universe, which in a way are unique experiments. For this reason, the development of more general theoretical concepts and the technological construction of the varying sensors, detectors, recorders, and computers have also become very general, aside from the industrial and commercial importance of these uniquely needed high-tech sensors. In many cases, the development of equipment for Astrophysics gives Physics new tools for use in everyday living, contributing and enhancing the lives of people. Particularly with high technology - born in research, and running through every stage of the manufacturing, sales promotion, and parts of the consumers and population needs at large. This book reports where S & T have evolved in its closeness this year. With its stars and exoplanets of different cosmic distances. Besides the significant contribution to human culture that comes from the study of space and time of our world, and the contribution of its pluralities of various other nations to the understanding of the concepts established by the immense current empire United Scientific. This capability for the conquest of the Universe represents the future of the world and of humanity currently threatened by the poverty in understanding of cultural and natural wealth, and by the potential abuses of people who dominate the economic power, and of course in providing justice to all, aiming at peace. Amen.

Astrophysics is the physical science dedicated to the study of the Universe. This book presents a panorama of the knowledge about our Universe obtained in the last few decades, at a pace much faster than in all the scientific history. Studies on the beginning of the cosmos, its ingredients and history, in addition to the understanding that by this narrative humans are an integral part of it, contribute and improve our daily life, as shown by the multiple cosmic arteries that power our civilization, formed mostly of very long-lived stars, whose energetic and chemical elements in their deaths are used by vegetable life to make carbohydrates: the food for animal and human bodies.

We conclude the book with a brief overview of the main topics of astrophysics and why the study of the Universe is essential to both enrich our daily experience and guarantee the future of humanity.

Milton Keynes UK
Ingram Content Group UK Ltd.
UKHW030908271124
451618UK00011B/341